Bibliographic information published by the German National Library:

The German National Library lists this publication in the National Bibliography; detailed bibliographic data are available on the Internet at http://dnb.dnb.de .

Imprint:

Copyright © 2016 GRIN Verlag, Open Publishing GmbH
Print and binding: Books on Demand GmbH, Norderstedt Germany
ISBN: 9783668265455

This book at GRIN:

http://www.grin.com/en/e-book/323584/behaviour-of-a-herbivore-plankton-conti-nuous-interaction-model

Kingsley Eshun Gyekye

Behaviour of a herbivore-plankton continuous interaction model

GRIN Publishing

GRIN - Your knowledge has value

Since its foundation in 1998, GRIN has specialized in publishing academic texts by students, college teachers and other academics as e-book and printed book. The website www.grin.com is an ideal platform for presenting term papers, final papers, scientific essays, dissertations and specialist books.

Visit us on the internet:

http://www.grin.com/

http://www.facebook.com/grincom

http://www.twitter.com/grin_com

Contents

HERBIVORE–PLANKTON INTERACTION MODEL

Kingsley Eshun Gyekye

Department of Mathematics and Computer Science

Lobachevsky State University, Nizhni Novgorod, Russia

Abstract

This paper investigates and analyzes the behaviour of a herbivore-plankton continuous model. Two of the equilibrium points are solved analytically while the third equilibrium point is solved with the help of Nullclines phase portrait. The model's equilibrium points stability and their ecological implications are analyzed and computer simulations are used to exhibit the characteristics of the model.

Keywords: Equilibrium Point, Stability, Herbivore-Plankton Model, Growth Rate.

1   Introduction

Population growth is one of the biological studies not easy to predict as it involves multiple variables some of which are almost impossible to determine. Engineers and researchers have used mathematical modeling and computer simulations to solve and predict many complex problems which would have been difficult to predict (Dym, 2004). Same can be said about the use of similar techniques for solving ecological problems. Modeling and qualitative analysis of population growth is one of the interesting areas in population ecology as it involves the application of discrete, continuous, linear and nonlinear differential equations. It is important in population ecology as it can help predict either increase or decline in population growth rate at any particular point in time. In the case of farming, modeling and analysis of plants can help farmers predict how well their crops will fare under different environmental conditions and even help them to predict future yields. It can also be used to predict whether a particular plant or animal species is on the verge of extinction Rockwood (2006).

2

In population ecology, single-species population is the simplest to model as it involves few parameters; a general model can be presented as

$$N(t + 1) = N(t) + B(t) - D(t) + I(t) - E(t), \ldots \ldots \ldots (1)$$

Where,

$B(t)$ is the birth rate at time t,

$D(t)$ is the death rate at time t,

$E(t)$ is the emigration rate at time t,

$I(t)$ is the immigration rate at time t,

$N(t)$ is the previous population size at time t,

$N(t + 1)$ is the current population size at time (t+1).

The equation (1) model can also be presented in a differential equation form as

$$\frac{dN}{dt} = rN, \ldots \ldots \ldots (2)$$

where r = (birth rate - death rate + immigration rate - emigration rate).

Population growth of such models increases exponentially especially when their habitat have the abundance of resources to support their numbers. In the case of plants; immigration, emigration and death rate is a small rate since the model does not include herbivores that feed on them.

Two-species population models in population ecology involve two discrete or continuous model of two different species either in competition for food or one killing the other for food. Examples of such models are the predator-prey model and the herbivore-plant model. The most popular of this type of model is the Lotka-Volterra predator-prey model.

In 1926, Lotka-Volterra developed a model that described the existence of a particular fish species predating on another fish species in the Adriatic Sea, explaining why there were not consistency in the level of fish catch in the Adriatic Sea (Murray, 2002).

3

We formulate the Lotka Volterra's predator-prey model by denoting;

$N(t)$ as the prey population at time t and

$P(t)$ as the predator population at time t,

then Volterra's predator-prey model is

$$\frac{dN}{dt} = (a - bP)N, \dots \dots (3)$$
$$\frac{dP}{dt} = (cN - d)P, \dots \dots (4)$$

where a, b, c and d are positive constants.

He made the following assumptions,

1. From equation (3), in the absence of predation, that is $(P = 0)$, the prey grows exponentially without restriction and the result is equivalent to equation (2).

2. In the case of the existence of predation then the prey growth rate is reduced by the $(-bP)$ term.

3. From equation (4), in the absence of prey that is $(N = 0)$, the population of the predator decrease exponentially due to the $(-dP)$ term.

4. Finally, with the existence of prey in equation (4) the population of the predator increases proportionally to the density of the prey.

2   Mathematical model

In reality, it is only poisonous plants or planktons that do not have herbivores to feed on and therefore models of single plant species without herbivores are best considered for studying purposes because even in a population model where there is coexistence of two species, it is difficult to say that, the existence of one may not affect the other as one may be a food for the other or might kill the other without being aware of the effect (Leah, 2005).

We will consider a model that is more precise in describing a population. Herbivores feed on plants or planktons. Herbivores feeding on planktons reduce the population of the planktons and that can even affect the reproduction cycle of the planktons. On the other hand, the existence of a large amount of planktons will boost the population of the herbivores.

We consider a herbivore-plankton interaction model version of the Lotka-Volterra predator-prey model cited in (Hirsch, Smale, and Devaney, 2013).

$$\frac{dx}{dt} = rx\left\{(k-x) - \frac{by}{c+x}\right\}, \dots \dots \dots (5)$$
$$\frac{dy}{dt} = dy\left\{\frac{x}{c+x} - ay\right\}, \dots \dots \dots (6)$$

where:

x: Represents plankton population,

y: Represents herbivore population,

a: Represents the minimum feeding rate needed by the herbivore to survive,

b: Represents the maximum feeding rate,

c: Represents half of the maximum feeding rate,

d: Represents herbivore growth rate when there are enough plankton to survive on,

k: Represents the assimilation efficiency of the plankton,

r: Represents plant growth rate when y_n is almost zero,

and a, b, c, d, k, r are all positive constants.

2.1 Solving for the equilibrium points

In this section, we solve for the equilibrium points of the algebraic equations (5) and (6) by equating them to zero. That is

$$\frac{dx}{dt} = 0 \text{ and } \frac{dy}{dt} = 0. \dots \dots \dots \dots (7)$$

First, we will solve for the solutions of equation (6) and substitute them into equation (5). By equating equation (6) to zero, we get a simple algebraic quadratic equation

5

$$dy \left\{ \frac{x}{c+x} - ay \right\} = 0. \ldots \ldots \ldots (8)$$

Therefore,

$$dy = 0 \text{ and } \left\{ \frac{x}{c+x} - ay \right\} = 0, \ldots \ldots \ldots (9)$$

which gives two solutions of equation (6) at $y = 0$ and $y = \frac{x}{a(c+x)}$.

We will only solve for the equilibrium point at $y = 0$ and use a graph to find the other positive equilibrium point in the first quadrant of the Cartesian plane.

We substitute $y = 0$ into equation (5) and the result is

$$rx\{(k - x)\} = 0. \ldots \ldots \ldots (10)$$

We get two linear equations $rx = 0$ and $(k - x) = 0$ from equation (10).

We get $x = 0 \text{ and } x = k$.

Therefore the points $(0,0) \text{ and } (k, 0)$ are two of the equilibrium points of the systems and there are others which would be investigated using Nullclines phase portrait.

2.2 Investigating the stability of the equilibrium points

We need to determine the stability of the two equilibrium points. First, we need to find the related eigenvalues of the linearization matrix at the equilibrium.

We let,

$$\frac{dx}{dt} = rx \left\{ (k - x) - \frac{by}{c+x} \right\} = G(x, y), \ldots \ldots \ldots (11)$$

and,

$$\frac{dy}{dt} = dy\left\{\frac{x}{c+x} - ay\right\} = H(x,y), \dots\dots\dots (12)$$

The next step is to solve for the eigenvalues. We use the formula, $determinant(A - \epsilon I) = 0$ where A is the linearization matrix at the equilibrium for the system of two equations, I is a unitary matrix and ϵ is eigenvalue.

We solve for A by using the Jacobian matrix,

$$J(x^*,y^*) = \begin{bmatrix} \dfrac{dG}{dx} & \dfrac{dG}{dy} \\ \dfrac{dH}{dx} & \dfrac{dH}{dy} \end{bmatrix} = A. \dots\dots\dots (13)$$

We then solve for the partial differentials of the equations (5) and (6), which gives

$$\frac{dG}{dx} = r\left\{k - 2x - \frac{bcy}{(c+x)^2}\right\}, \dots\dots\dots (14)$$

$$\frac{dG}{dy} = -\frac{rbx}{c+x}, \dots\dots\dots (15)$$

$$\frac{dH}{dx} = \frac{cdy}{(c+x)^2}, \dots\dots\dots (16)$$

$$\frac{dH}{dy} = \frac{dx}{c+x} - 2ady, \dots\dots\dots (17)$$

Substituting the partial differentials into equation (13), we get

$$J(x^*,y^*) = \begin{bmatrix} r\left\{k - 2x^* - \dfrac{bcy}{(c+x)^2}\right\} & -\dfrac{rbx}{c+x} \\ \dfrac{cdy}{(c+x)^2} & \dfrac{dx}{c+x} - 2ady \end{bmatrix}. \dots\dots\dots (18)$$

This implies that using the equilibrium point (0,0), we get

$$J(0,0) = \begin{bmatrix} rk & 0 \\ 0 & 0 \end{bmatrix} \quad \dots \dots \dots (19)$$

We proceed to calculate for the eigenvalues of the equilibrium point $(0, 0)$,

$$\det[J(0,0) - I \in] = 0, \dots \dots \dots (20)$$

$$\det \begin{bmatrix} rk-\in & 0 \\ 0 & -\in \end{bmatrix} = 0, \dots \dots \dots (21)$$

$$-\in (rk-\in) = 0, \dots \dots \dots (22)$$

$$\in = 0 \;\; and \;\; \in = rk, \dots \dots \dots (23)$$

The critical point $(0,0)$ is unstable node since $rk > 0$.

We consider the second equilibrium point$(k, 0)$ and investigate its stability. We follow the same the procedure used to investigate the stability of point $(0, 0)$.

For equilibrium point (k, 0), it follows that

$$J(k,0) = \begin{bmatrix} -rk & -\dfrac{rbk}{c+k} \\ 0 & \dfrac{dk}{c+k} \end{bmatrix}, \dots \dots \dots (24)$$

$$\det \begin{bmatrix} -rk-\in & -\dfrac{rbk}{c+k} \\ 0 & \dfrac{dk}{c+k}-\in \end{bmatrix} = 0, \dots \dots \dots (25)$$

$$-(rk+\in)\left(\dfrac{dk}{c+k}-\in\right) = 0, \dots \dots \dots (26)$$

$$\in = -rk \;\; and \;\; \in = \dfrac{dk}{c+k}. \dots \dots \dots (27)$$

The critical point $(k, 0)$ is saddle since $\frac{dk}{c+k} > 0$.

2.3   Computer simulations and the qualitative analysis

In order to visualize and analyze the behaviour of the herbivore-plankton model, a Matlab application was used.

Nullclines phase portraits of the herbivore-plankton model

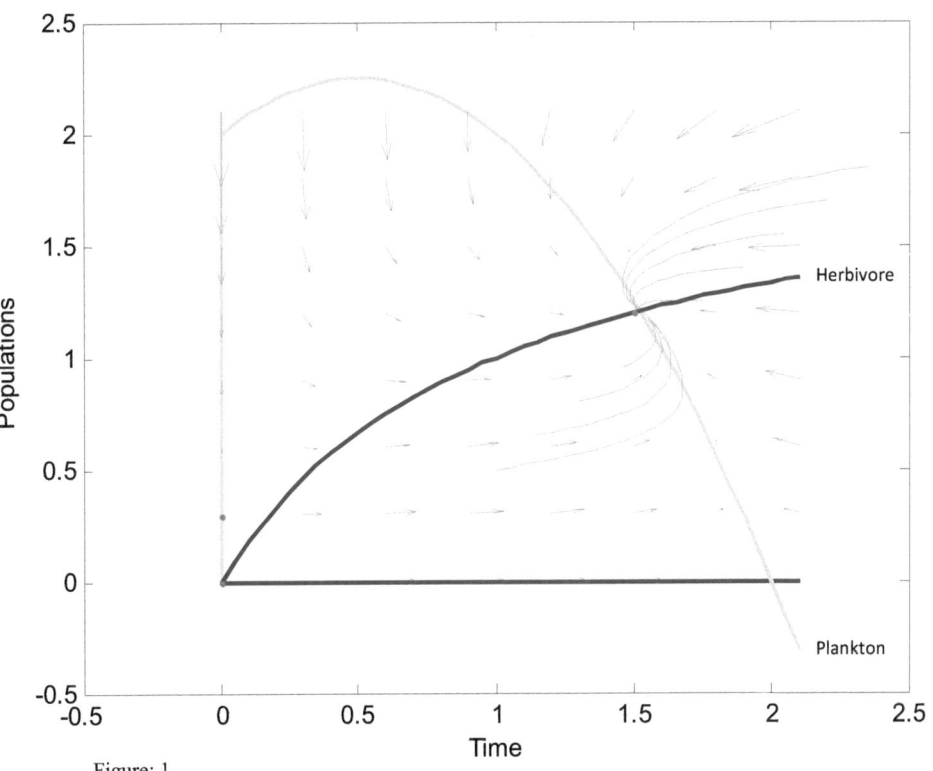

Figure: 1

Previously, we solved for two equilibrium points of the systems and we realized that the equilibrium point (0, 0) is unstable and the point (k, 0) is saddle. We can deduce from the (Figure 5.1) that, the system has a third positive equilibrium which is a stable equilibrium point even though this point depends on the choice of parameter values of the system. In this case with the parameters values r=1, k=2, a=0.5, b=1, c=1, d=1, the system has another equilibrium point at approximately (1.5212,1.2066) which is stable.

9

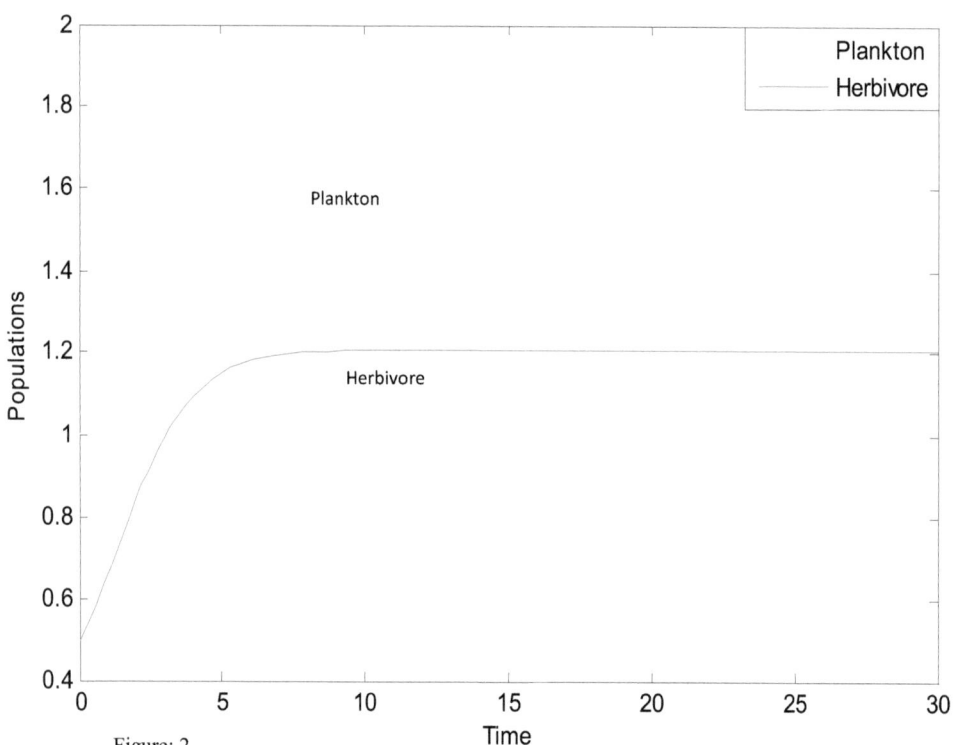

Figure: 2

Using the following parameters r=1, k=2, a=0.5, b=1, c=1 and d=1.

We can observe from the (Figure 5.2) that the planktons increased exponentially for the first 2 years but declined gradually till it attained stabilization after 8 years. The herbivores also increased exponentially for the first 5 years but stabilized after that.

Logistic graph of the herbivore-plankton model

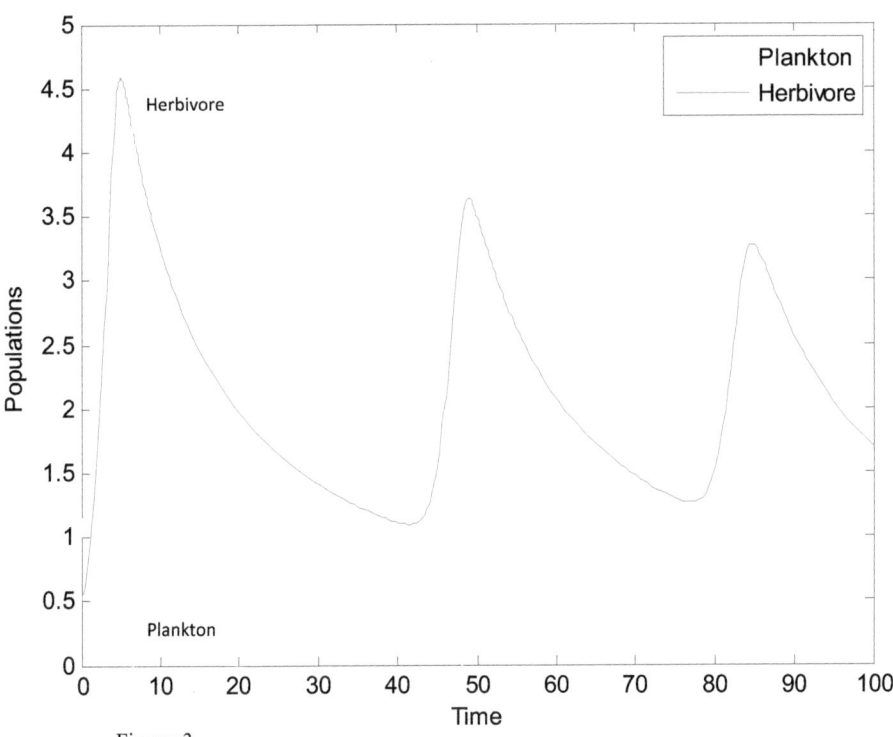

Figure: 3

Using the following parameters r=1, k=2, a=0.02, b=1, c=1 and d=1.

We observe that both systems do not exhibit stable behaviour but rather exhibit fractal like characteristics.

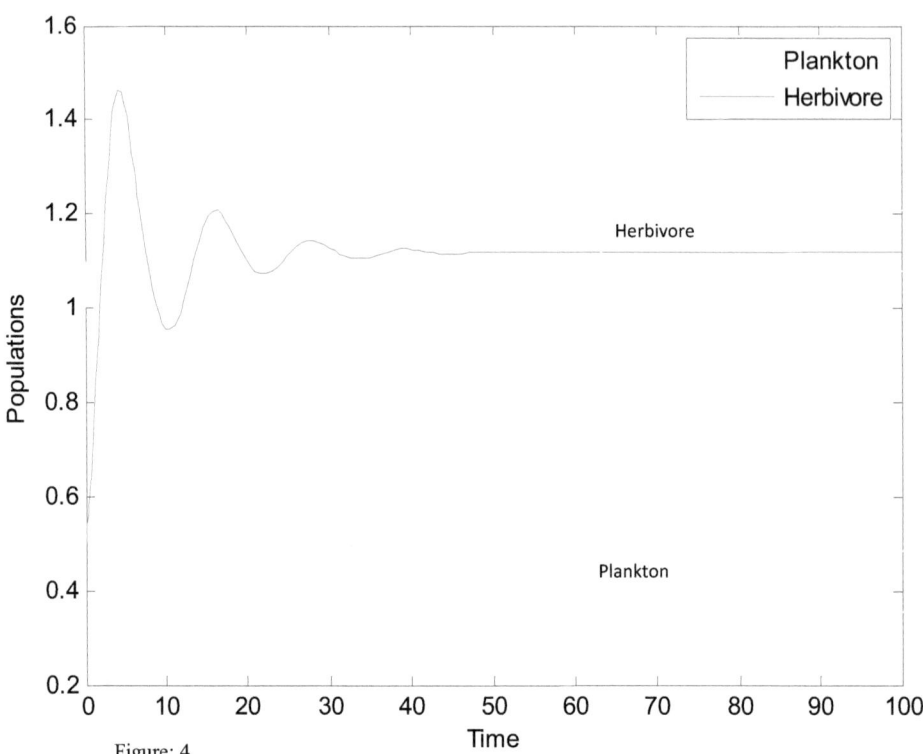

Logistic graph of the herbivore-plankton model

Figure: 4

Using the following parameters r=1, k=2, a=0.25, b=2, c=1 and d=1.

We observe that both systems exhibit unstable behaviour but tend to stabilize after time 40.

3   CONCLUSIONS

In a general case, plant-herbivore interactions are not easy to predict since their characteristics depend on various parameters. From (Figure 5.2), using the following parameters r=1, k=2, a=0.5, b=1, c=1 and d=1. We can observe from the (Figure 5.2) that the planktons increased exponentially for the first 2 years but declined gradually till it attained stabilization after 8 years. The herbivores also increased exponentially for the first 5 years but stabilized after that. That is, with the right value of parameters, both plankton and herbivore increased and stabilized over years. One of the possible interpretations is that, with a large amount of plankton and a few number of herbivore to consume, the population of plankton will still be increasing since the consumption does not significantly affect their growth but as the population of the herbivore increases, at a point, the growth of the plankton decreases because of the significant amount of consumption by the herbivore. After some years, both plankton and herbivore population stabilizes making it good for their long existence in the ecosystem.

REFERENCES

1. Dym,C. L. (2004). Principles of Mathematical Modeling, 1st Edition. Publisher: Elsevier Academic Press, 2004.

2. Murray, J. D. (2002). Mathematical Biology I, An Introduction Third Edition. Publisher: Springer-Verlag New York Berlin Heidelberg.

3. Leah, E. K. (2005). Mathematical Models in Biology. Publisher: Random House, New York, NY, 1988.

4. Rockwood, L. L. (2006) Introduction to Population Ecology. Publisher: Blackwell Publishing Ltd., 9600 Garsington Road, Oxford OX4 2DQ, UK, 2006.

5. Hirsch, M. W, Smale, S. and Devaney, R. L. (2013). Differential Equations, Dynamical Systems, and an Introduction to Chaos. Publisher: Academic Press, 2013.

APPENDIX

The following Matlab code were used for the visualization of the graphs

Code 1

Matlab code for (Figure 1)

```
function [ dx ] = Herbivoreplant( t, x )
global r a b c d k;
dx = zeros(2,1);
dx(1)=r.*x(1).*((k-x(1))-b.*x(2)./(c+x(1)));
dx(2)=d.*x(2).*(x(1)./(c+x(1))-a.*x(2));
end


global r k a b c d;
r=1; k=2; a=0.5; b=1; c=1; d=1;
Tmax=30;
x0=1;
y0=0.5;
x3=0:0.05:2.1;
ypx1=(k-x3).*(c+x3)/b;
ypy1=x3./(a*(1+x3));
plot(x3,ypx1,'Color','green','Linewidth',2);
hold on;
plot(x3,ypy1,'Color','blue','Linewidth',2);
hold on;

for i=1:1:10
    ypx2=0*x3;
    plot(x3,ypx2,'Color','blue','Linewidth',2);
    plot(ypx2,x3,'Color','green','Linewidth',2);
    hold on;
```

end

```
[x1, x2] = meshgrid(0:0.3:2.3, 0:0.3:2.3);
x1dot = r*x1.*((k-x1)-b*x2./(c+x1));
x2dot = d*x2.*(x1./(c+x1)-a*x2);
quiver(x1,x2,x1dot, x2dot,'Color','red');
hold on;

for i=1:1:10
    [t,x]=ode45(@Herbivoreplant,[0 Tmax],[x0 y0]);
    plot(x(:,1),(x(:,2)),'Color','red');
    hold on;
    x0=x0+0.15;
    y0=y0+0.15;
end
hold off;
```

Code 2

```
Matlab code for (Figure 2)
global r k a b c d;
r=1; k=2; a=0.5; b=1; c=1; d=1;
Tmax=30;
x0=1;
y0=0.5;
[t,x]=ode45(@Herbivoreplant,[0 Tmax],[x0 y0]);
plot(t,x(:,1),'color','green');
hold on;
plot(t,x(:,2),'color','blue');
```

Code 3

Matlab code for (Figure 3)

```
global r k a b c d;
r=1;  k=2;  a=0.02;
b=1;  c=1;  d=1;
Tmax=100;
x0=1;
y0=0.5;
[t,x]=ode45(@Herbivoreplant,[0 Tmax],[x0 y0]);
plot(t,x(:,1),'color','green');
hold on;
plot(t,x(:,2),'color','blue');
```

Code 4

Matlab code for (Figure 4)

```
global r k a b c d;
r=1;  k=2;  a=0.25; b=2;
c=1;  d=1;
Tmax=100;
x0=1;
y0=0.5;
[t,x]=ode45(@Herbivoreplant,[0 Tmax],[x0 y0]);
plot(t,x(:,1),'color','green');
hold on;
plot(t,x(:,2),'color','blue');
```